As all of the goods could easily break, she can't even make a tiny mistake.

Packing the cups takes a lot of care —
and she has to check that they're all there.

The trays are then put in a stack,
and tied up tightly in one big pack.

Polly packs bowls in sets of three,
in plastic trays so they're easy to see.

Then she packs these inside a box,
to keep them safe from bangs and knocks.

That only leaves the delicate plates, which Polly arranges in stacks of eight.

She puts soft padding between each stack —
she wants to be sure that they won't crack.

At last Polly's finished her work for the day, and all of the orders can be sent away.

Written by Calvin Irons
Illustrated and designed by Kim Roberts

The author would like to thank Janet Hillman for her contribution to this project.

© 1993 Mimosa Publications Pty. Limited

All rights reserved
97 96 95 94 93
8 7 6 5 4 3 2 1

Printed in Australia
ISBN 0 7327 1158 4

Published in the United States of America by
MIMOSA PUBLICATIONS
P.O. Box 26609
San Francisco CA 94126
(800) 443 7389

Published in the United Kingdom by
KINGSCOURT PUBLISHING
P.O. Box 1427
London W6 9BR

Published in Canada by
GINN PUBLISHING CANADA INC.
3771 Victoria Park Avenue
Scarborough
Ontario M1W 2P9

Published in Australia by
MIMOSA PUBLICATIONS
8 Yarra Street
Hawthorn
Victoria 3122

Level 3 Operations — multiplication